To Robert   Xm

From

DIG YOUR OWN GOLD

Written and Compiled By

Ellen Gallup Genta

Hawkes Publishing Inc.
3775 South 500 West
(P.O.Box 15711)
Salt Lake City, Utah 84115
Phone (801) 262-5555

ISBN # 0-89036-062-6

DEDICATION

"DIG YOUR OWN GOLD" is dedicated to my Father, Ellis Strong Gallup, the little salesman who turned failure into *SUCCESS*. Many of the following bits of *WISDOM* were a part of his humor, wit and life.

Also to my Mother, Rosa Love Gallup, who was patient enough to put together "My Dad's Scrapbook."

FOREWORD

"DIG YOUR OWN GOLD" is a collection of bits of *WISDOM* and fun. They are taken from a cherished scrapbook and from other valuable meditations that will contribute to any circumstance or profession.

"Knowledge is useless unless it is harnessed to *YOUR* accomplishment." Take up these tools of *WISDOM* and let them help you *"DIG YOUR OWN GOLD."*

Written and compiled by
Ellen Gallup Genta

INTRODUCTION

Like you, I am the perennial prospector searching for the "pot-of-gold" at the end of the rainbow and enjoying every moment of it. "Finding gold is a digging operation." For many years I have been digging in the mine called *BUSINESS* and have been working its two largest veins. Namely, buying and selling. Luck has been with me in this operation for I have found rich deposits of ore in the friends I have made.

The vein I am presently digging in is a rich strike. Its walls are a high standard college book store and the precious ores are the students and the choice people involved therein. A significantly exciting phase of my digging operation is the training of some of these precious ores. Preparing the training lessons inspired me to dig out my Dad's scrapbook and give "To *YOU*" these bits of Wisdom. They are the Mother-lode. Should you add only one, each day, to your miner's pouch, you will be more likely to reach your "pot-of-gold."

My Dad's scrapbook is a collection of quotable quotes, poems and free advise from the company he worked for. He was a door to door doer. This type of selling is the testing ground of all salesmanship. It is often a lonesome profession, full of impolite words and criptic remarks that goad you to give up. Should you become successful, you will have won a special medal from the school of hard knocks and slammed doors. My dad earned this medal.

He was a man of small stature, often ridiculed and sold short by many. His earlier working days were spent on a farm. It was well known that he was slow, easy going, full of good humor and wit. He was always the last one to get the cows milked and was thus rated as a poor farmer. Late in life he made a change in his occupation and became a salesman, a successful one.

The Gold he extracted from this mine was infinitely more precious than opals or diamonds. He accomplished this by having the courage and determination to go back after slammed doors and empathetic "No's." He carried with this peddlers case a bucket of confidence capped with friendliness. Liking people, respecting his company and his job were the grooves in the handle of his shovel. The claim he worked was "your town and my town." Though he passed by the fountain of youth and was considered to be going down hill, he vowed the choice was fortunate. He always said, "Better late, than never, but better never late."

He taught me to prospect at a very young age. I remember a day, coming in from the pasture after hearding the cows. Finding no chores left to do and no neighbor children to play with, I approached my Dad.

"What can I do? The cows are in the pasture across the canal. There is nothing to do," I pestered.

"Take a shovel, go out in the garden and dig me a hole two feet wide and two feet deep," was his laconic reply.

After digging the ordered hollow, I was back. "Now what?" I asked again.

'Fill it up again, at least your hands will be busy," was his long remembered lesson.

I found soon enough that a mine produced in accordance to my persistance and knowledge. The upshot of this was; if I seemed unduly quiet at long intervals, I was frequently found up in a favorite tree with a book in my hand. This was my "tree of knowledge." The lesson from my Dad taught me to search in more fertile and rich claims than a garden. I soon learned there were mountains of precious deposits to be *WORKED.* In later years, one such rewarding vein was my Dad's scrapbook.

As a salesman, I don't assume that he always sold the highest quota in his product but I do know he sold his *PERSONALITY.* Very often, after ten years, many people still say, "Oh I remember your Father, he's the little fellow who always had a smile and a story. We enjoyed having him come to our home." Few salesmen. or even few people can have such a long remembered epitaph.

During these years, he was sent stimulating bulletins and letters from the company he worked with. They were filled with such interesting and enthusiastic ideas that my Mother patiently placed them in a scrapbook.

Because my work was also involved in selling and working with people, my Mother gave the book to me. Since my Mother and Father are no longer with me, it remains one of my most treasured possessions.

DIG YOUR OWN GOLD is a series of items from this book, along with many others from which I have received inspiration. At no time do I take credit for the originality of these quotes. I have used the authors names if known. These golden thoughts I will pass along to you, giving a very special thanks to those who have written them. I have not intentionally given any special consideration to any particular person or company. Any mention of my Dad's company is known to them. I have used his material to show what a man of little stature, at any age, can accomplish.

Since any walk of life or profession makes each of us a salesman, requiring that we sell our *PERSONALITY*, our *ENTHUSIASM*, our ambition or our abilities to someone, sometime; I give to you this combination of facts, fun and free advise. How you use these tools to promote your mine depends on *YOU*. Many failures can be traced to insufficient development. Two frequently, hate, envy, optimism and laziness blind the prospector to the most priceless nuggets. *DO NOT LET THE ROCK OF FAILURE DULL YOUR SHOVEL.* Let your ability open the shaft of *SELLING;* operate it by *WORK,* energize it with *EXPERIENCE*—by climbing the *LADDER OF CHARACTER* which will yield *SUCCESS AND! FAILURE OR? HAPPINESS AND!!!* -

Written and compiled by
Ellen Gallup Genta

ACKNOWLEDGEMENTS

The Fuller Brush Company .... Hartford, Conn.
Richard L. Evans Quote Book .... Publishers Press
Joy In My Life .... Garrett E. Case
The Dartnell Corporation .... Chicago, Ill.

Sincere appreciation to my sister, Grace Forsyth, Betsy Matthews, Dawna Derr, Carmen Garner, Weldon Steiner, Blaine M. Yorgason and Ronald K. Messer.

CONTENTS

SELLING

PART 1

SELLING OR?

"Some people fall into fortune, but nobody falls into *SUCCESS.*" A greater number of us *WORK* for it. Where-ever you *WORK*, or whatever your *PROFESSION*, read and digest this "gold mine" on salesmanship motivation. Here are tools of friendliness, courtesy and service that will give you confidence whether you are *SELLING OR?* These same tools were once in the hands of the "Greatest *SALESMAN* On Earth."

"Selling is a digging operation," so lets dig. Even if someone gives you a "gold mine," you have to dig the gold. These ores are down to earth, practical common sense gems with many facets that can sparkle their way into your days *WORK.*

★★★★★★★★★★★★★★★★★★

Diamond Jim Brady once said, "No matter what you're selling, the most important thing you're selling is *YOURSELF.*

★★★★★★★★★★★★★★★★★★

Your business is like a wheelbarrow;
it stands still unless it is pushed.

When you sell a thing, it is because the buyer has a need. You are doing something constructive. Never apologize for filling a need.

★★★★★★★★★★★★★★★★★★

You can't drive a nail with a sponge,
No matter how hard you sock it.
And you can't sell merchandise from a rocking chair,
No matter how hard you rock it . . . Scrapbook

★★★★★★★★★★★★★★★★★★

When a goat starts backing up, he is getting ready to do something. When a *SALESMAN* starts backing up, he's through . . . Cox

★★★★★★★★★★★★★★★★★★

Selling would be crowded if boys could do it.

★★★★★★★★★★★★★★★★★★

Don't forget that loyalty is one of the greatest virtues in selling.

Don't be too aggressive, there's a happy medium between high prssure and a flat tire.

Don't take knowledge for granted. Assume that your customer doesn't know all the advantages of quality—tell her.

Don't forget the word *YOU* in talking to a prospect.

. . . Scrapbook

★★★★★★★★★★★★★★★★★★

FORMULA FOR HANDLING PEOPLE
1. Listen to the other person's story.
2. Listen to the other person's full story.
3. Listen to the other person's full story first.

. . . General George C. Marshall

TEN LITTLE CUSTOMERS

TEN LITTLE CUSTOMERS, all of them were mine,
   I mispelled the name of one, then there were nine.
NINE LITTLE CUSTOMERS, sober and sedate,
   I kidded one too much, then there were eight.
EIGHT LITTLE CUSTOMERS, thinking lots of heaven,
   I swore in front of one, then there were seven.
SEVEN LITTLE CUSTOMERS wouldn't stand for tricks,
   I broke a date with one, then there were six.
SIX LITTLE CUSTOMERS, all of them alive,
   I Yawned in front of one, then there were five.
FIVE LITTLE CUSTOMERS, full of wighty lore,
   I tried to pull a bluff, then there were four.
FOUR LITTLE CUSTOMERS, buying brushes from me,
   I got an order wrong, then there were three.
THREE LITTLE CUSTOMERS, getting mighty few,
   I knocked a competitor, then there were two.
TWO LITTLE CUSTOMERS, I had 'em on the run,
   Talked of one behind his back, then there was one.
ONE LITTLE CUSTOMER, wasn't any fun,
   Too much *WORK* to call on him, then there was none.
                                          . . . Scrapbook

★★★★★★★★★★★★★★★★★★

The smart cookie sells more than the *WISE* cracker.
                              . . . The Dartnell Corporation

★★★★★★★★★★★★★★★★★★

CAN YOU SPOT A GOOD SALESMAN?
   He will have a steady eye, a steady nerve, a steady tongue
and steady habits. He will keep his temper and his *FRIENDS*.
   A good *SALESMAN* (and a good man) can be spotted by:
   HIS HUMOR; humor is at the bottom, a balance wheel and
neither a salesman nor a watch are much good without it.
   HIS TEMPER: Keep your temper, no one else wants it. A
man's temper is like steel, neither are of value when they have
lost their temper.

HIS ENERGY: Energy is a sure sign of physical fitness and is as the hiss of steam to a locomotive.
FRIENDLINESS: Selling is human relationship, basic and sincere friendship. Friendliness is the most precious thing anyone can have. It is sometimes as fragile as an orchid, as precious as gold, as hard to find, as powerful as a great turbine, as wonderful as *YOU* and as hard to keep.

. . . Scrapbook (entwined with my own thoughts).

The travailing *SALESMAN* takes his troubles from door to door.

Buying cheap merchandise to save money is like stopping the clock to save time. Selling cheap merchandise would also stop the clock, along with your sales.

AN OLD NOSTALGIC STORY
OF A SELLING DEMONSTRATION

"How much am I bid for this fiddle?" asked the auctioneer. "One dollar? Who'll bid two? Who'll bid three?" No one wanted the old fiddle and the auctioneer was about to sell it for two dollars, when an old man stepped to the platform.

"One moment," he said, and took up the fiddle. He dusted it off carefully. He tuned it up. He put it to his chin and began to play a song of long ago.

The music touched the hearts of cold hearted bidders. As he finished, the old man patted the fiddle and put it on the block, wiped a tear from his eye and hobbled from the platform.

The autioneer broke the silence, "What am I bid?" I'll bid ten dollars," said one. "I'll bid twenty-five," said another. A third man stood up and cried out, "I'll bid one hundred dollars for the old man's fiddle."

What had happened to suddenly make the old fiddle sell? *THE OLD MAN HAD PUT SOME LIFE INTO IT*, merely by handling and demonstrating it with *LOVE*.

Star *SALESMEN* don't get discouraged—they don't have time.

Don't look at the obstacles, look at the possibilities.

★★★★★★★★★★★★★★★★★

Calvin Coolidge said, "Modern business could neither have been created nor can it be maintained without creative selling. It is not enough that goods be made; a demand must be made for them also.

★★★★★★★★★★★★★★★★★

At one time the Navy Department released a simple statement well worth observing: Loose words may lose ships. Beware of inquisitive friends. Do you know who is listening? Careless talk costs lives. Walls have ears. Do not repeat rumours. Gossip is faster than radio. Keep your own counsel. Idle words may reveal vital information.

... Scrapbook

Keep this motto on your shoulder. Don't be dubbed as a fast talking salesman. It has been said of me that, "She could sell an eskimo an igloo." My experience has taught me to greet my potential customers with candor and friendliness. Determine his needs and display the saleable product with ease and gentle methods but truthful sales talk. Most people judge selling methods by the way we would like them used on ourselves. As for myself, I think that high pressure, pusher selling never was any good. I am confident there is an intensive over-emphasis in our selling to Mr. America today. Would it not be more worth while to dig into the mine of selling and come up with methods to meet the wants and the needs more effectively. If you want a *satisfied customer,* you had better dig up more pleasing and constructive suggestions in demonstrating your ability and product.

To me, there are three kinds of customers; one who is calm, one who likes to crow and there is amicable Friendly Freddie.

A good salesman must have the intuition to converse fluently at the right time. With the clam, I advise; Do not use too much pressure, just enough to draw him out of his shell. The crow, "Let him crow." Before he gets through he will have undoubtedly convinced himself that he needs your product, with only a little touch of positive approach from you.

As for Friendly Freddie, you can enjoy your selling job here by showing genuine pleasure and interest in his wants and expectations. A wise salesman can do this without the use of "idle words" or "careless talk." The Friendly Freddies can give you an opportunity to "dig into new vistas of mutual good-will and challenges. It is a joy to serve *HIM.* If you want to sell an eskimo an igloo, meet your customer with a smile, a real open smile. A smile can melt ice and will also increase your *"FACE VALUE."*

★★★★★★★★★★★★★★★★

A FINE QUARTETTE

A little bit of *QUALITY* will always make 'em smile,
   A little bit of *COURTESY* will bring 'em back a mile.
A little bit of *FRIENDLINESS* will tackle 'em tis plain.
   And a little bit of *SERVICE WILL* bring 'em back again.
                                             . . . Scrapbook

★★★★★★★★★★★★★★★★★★★

He who has the truth never need fear the want of persuasion
on his *TONGUE* . . . Ruskin

★★★★★★★★★★★★★★★★★★★

A well dressed *SALESMAN* wears the corners of his mouth up
                                    . . . The Dartnell Corporation

★★★★★★★★★★★★★★★★★★★

A SUCCESSFUL SALESMAN MUST HAVE:
   The curiosity of a cat.   The tenacity of a bulldog.   The
determination of a taxicab driver.   The diplomacy of a wayward
husband.   The patience of a self-sacrificing wife.   The
enthusiasm of a flapper. The friendliness of a child. The good
humor of an idiot. The simplicity of a jackass. The assurance
of a college boy. The tireless energy of a collector of past due
bills.   . . . Scrapbook

★★★★★★★★★★★★★★★★★★★

   Most Salesmen do what is expected; the *SUCCESSFUL
SALESMAN* does a bit more.
                                    . . .The Dartnell Corporation.

Being average puts one as close to the bottom as to the top.

A survey once made to determine why sales people failed: thirty seven percent failed because of discouragement—thirty seven percent failed because of lack of industry—twelve percent did not follow instructions and eight percent because of lack of knowledge.

★★★★★★★★★★★★★★★★★

ALIBIS, what are they anyway? An alibi is only a piece of dough that started out to be a biscuit and ended up as a pancake. Alibis are sometimes called weather reports, but never used by good salesmen. Good salesmen make no alibis

. . . Scrapbook

★★★★★★★★★★★★★★★★★

A good man gives orders to himself. He then prepares to carry them out expertly and efficiently.

★★★★★★★★★★★★★★★★★

FROM EDGAR A. GUEST:
Customers are people who, Long remember what you do. If a sneer your face has crossed, As a salesman you have lost?
If you think you have the right to, Pick the ones you'll be polite to,
Soon you'll find to your dismay, You've let your business get away.
NEVER MIND YOUR WHIMS AND FANCIES, GIVE YOUR BEST, AND TAKE NO CHANCES.

Are you a butterfly? Do you flit from one position to another? Pick a reputable firm and light long enough to give all angles an earnest try.

. . . E.G.G.

★★★★★★★★★★★★★★★★★

If I possessed a shop or store,
  I'd drive the grouches off my floor.
I'd never let some gloomy guy,
  Offend the folks who came to buy;
I'd never keep a boy or clerk,
  With mental toothache at his *WORK*,
Nor let a man who draws my pay,
  Drive customers of mine away.

. . . Edgar A. Guest

★★★★★★★★★★★★★★★★★

MY BEATITUDES ON CUSTOMER RELATIONS:
  Blessed is he who behaves toward his customer or friends with respect and honor, Who does not complain and feel sorry for himself while making a sale, Who does not try to bluff people into buying, Who does not criticize his competitors product, Who does drive a "fair bargain" and gives his customer "his due," Who remembers to be thankful, Yea, Blessed is he who renders a service to mankind, whether selling, sowing or reaping the results.

★★★★★★★★★★★★★★★★★

The old saying that opportunity knocks but once is false. Opportunity keeps regular office hours from twelve midnight to twelve midnight, every day, every year. Opportunity will not call on you. You have to go up and meet it. Do you think any man ever happened to get to the top of the mountain. When you see a man there you will know he climbed there. Opportunity waits for no one.

. . . Scrapbook

★★★★★★★★★★★★★★★★★★

HAVE YOU READ:"THE AUTOPSY OF A LOST SALE?"

I lost to my customer because I made exaggerated claims about my product.

I spent all my time on the person easiest to reach instead of the important man in the company.

I let the buyer scare me.

I kept my big mouth open too long after I made the sale.

I didn't cover my sales presentation thoroughly.

I didn't know enough about my prospects business.

I didn't sell myself thoroughly enough.

I got into an argument with the prospect and won it.

I didn't know enough about my own product.

I spent too much of my time knocking my competitor.

I didn't generate enough desire for my product.

I didn't plan my presentation carefully enough.

I lost the sale to a competitor with a lower price because I couldn't tell why my product was worth more.

. . . Scrapbook

APPLY THIS TO YOUR DAILY ACTIVITIES.

This autopsy could be yours. Read it, study it and apply it. Don't let anyone put this headstone of *"LOST SALE"* on your grave of defeat. Abraham Lincoln said, "My great concern is not whether you have failed, but whether you are content with your *FAILURE.* Don't be content with "Bad Day Blues." Look up your Sure-Fire-Sales batting average and let it give you a *"LIFT."*

★★★★★★★★★★★★★★★★★★

TEN COMMANDMENTS OF HUMAN RELATIONS

1. Speak to people. There is nothing as nice as a cheerful word of greeting.
2. Smile at people. It takes 72 muscles to frown, only 14 to smile.
3. Call people by name. It is music to one's ear.
4. Be friendly and helpful. To have friends, be friendly.
5. Be cordial. Speak and act as if everything you do were a real genuine pleasure.
6. Be genuinely interested in people. You can like everybody if you try.
7. Be generous with praise—cautious with criticism.
8. Be considerate with the feelings of others.
9. Be thoughtful of the opinions of others. There are three sides to a story-yours-the other fellows-and the right one.
10. Be alert to give service. What counts most in life is what we do for others.

. . . Scrapbook

A WISE author gave these commandments to a world of industrious people. Put your mind in gear; spread these gospel tools wherever you sell your achievements and you will joyously become acquainted with the principle of *WORK*.

Work

PART II

WORK

WORK is the greatest remedy available for both mental and physical affliction . . . *TRY IT!*

★★★★★★★★★★★★★★★★★★

If disappointments come—*WORK*.
When faith falters—*WORK*.
When dreams are shattered and hope seems dead—*WORK*.
WORK as if your life were in peril—*IT IS!*
No matter what ails you—*WORK*.
If you are burdened with unfair responsibilities—*WORK*.
If you are poor—*WORK*.
If you are rich, continue to *WORK*.
If you are happy, keep *WORKING* faithfully.
WORK WITH FAITH.

. . . Scrapbook

★★★★★★★★★★★★★★★★★★

Idleness is the devil's workshop.

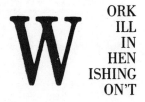

W ORK
ILL
IN
HEN
ISHING
ON'T

★★★★★★★★★★★★★★★★★

Get to hate *WORK* with such a passion that it doesn't pile up on you.

... Jerry Kernion

★★★★★★★★★★★★★★★★★

Footprints on the sands of time are not made by sitting down.

★★★★★★★★★★★★★★★★★

If the devil finds a man idle, he will put him to *WORK*.

★★★★★★★★★★★★★★★★★

The Lord gave us two ends, one to sit on and the other to think with. A man's *SUCCESS* depends upon which he uses most. It is a case of heads you win and tails you loose. *TAKE YOUR CHOICE, IT'S your challenge!*

★★★★★★★★★★★★★★★★★

The most valuable part of you is your head. If you do not use what it contains, the rest of your body will not amount to much.

The secret of life is not to do what one likes, but to like what one does.

★★★★★★★★★★★★★★★★★

When you have something to do, think it through, then *DO IT*. The results of a job economically planned and efficiently executed brings compensations in self satisfaction that can't be compared to any justified *CASH RECEIPTS.*

★★★★★★★★★★★★★★★★★

Nothing is so empty as a day without a plan. He who fails to *PLAN*, plans to *FAIL.*

★★★★★★★★★★★★★★★★★

If we don't try, we don't do; if we don't do, what are we on earth for.

. . . From Shenendoah.

★★★★★★★★★★★★★★★★★

Hard *WORK* never killed a man, but it sure has scared a lot.

★★★★★★★★★★★★★★★★★

Every man's task is his life preserver.

. . . George B. Emerson

★★★★★★★★★★★★★★★★★

THE FOUR LITTLE DEVILS

There are four little devils that wander about
    And camp on a poor salesman's trail.
They never think much of the fellow that wins,
    But they're strong on the fellows that fail.
They come well equipped to give you a fight,
    With harpoon and pitchfork and cane.
They cuddle up near, say nice things in your ear,
    But they're devils-look-out-just- the same.

The first little devil commences his work,
    As soon as it's getting-up time.
He wispers, "Roll over, go back to sleep,
    You can't see the buyer till nine."
And he tries to convince you that you're but a fool
To go out and hustle and strain;
He's a devil-look-out-just the same.

The second red devil unloosens his tongue,
    Before you are close to half through.
He says, "Go to dinner, you're working too hard,
    There's no use from eleven to two.
If you listen to him, you're fooling yourself
    And helping competitors gain.
He carries a smile and tongue full of guile,
    He's a devil-look-out-just the same.

The third little devil is oily and suave,
    He bides his good time to commence.
He waits for the time when the orders come slow
    And thinks that you won't take offense,
When he whispers so sweetly along about four,
    Let's knock off, Let's call it a game.
It may sound awfully nice to be through in a trice.
    He's a devil-look-out-just the same.

The fourth little devil's the worst of all
    And he smiles in a devilish way,
When he finally tells you that Saturday's yours;
    Should not work, just a plain loafing day.
And he knows very well if you listen to him,
    That you're bound to loose-cannot gain.
It's the easiest way but never will pay.
    He's a devil-look-out-just the same.

These four little devils, they sit on the fence,
    When it comes to reckoning day.
They laugh and they dance in devilish glee
    At the price they have forced you to pay.
When tempatation points to the easiest way,
    Take the hard road that leads up to fame.
The thing that's alluring can ne'er be enduring,
    THEY'RE DEVILS-LOOK-OUT-JUST THE SAME.

                                    . . . Scrapbook

★★★★★★★★★★★★★★★★★

I never did a days *WORK* in my life, it was all fun. I never
did anything by accident, nor did any of my inventions come
by accident.

                                    . . . Thomas A. Edison

★★★★★★★★★★★★★★★★★

No matter how lofty your ambitions may seem, do not
ridicule or stifle them. Live them, dream them. Keep on and
ambitions will come true.

                                    . . . Larabee

★★★★★★★★★★★★★★★★★

A plan 75% perfect but followed 100% is more valuable than a plan 100% perfect but followed 75%.

Without ambition one starts nothing. Without *WORK* one finishes nothing. The prize will not be given to you.

You have to win it. A man who knows how will always have a job. The man who knows why will always be boss.

. . . Emerson

The heights by great men reached and kept
  Were not attained by sudden flight,
But they, while their companions slept,
  Were toiling upward in the night.

. . . Longfellow

They that *WORK* not, cannot *PRAY*.   . . . Dwight

If you do your *WORK* with your whole heart, you will *SUCCEED* because there's so little competition.

When you make your job important, it will make you important.

You can't keep your eye on the ball and the clock at the same time.

★★★★★★★★★★★★★★★★★

WORK, don't worry, let God watch over the world and get a good nights sleep.

. . . Edgar A. Guest

★★★★★★★★★★★★★★★★★

WORRY, like a rocking chair, keeps you busy but gets you nowhere.

. . . Bradford Thomas

★★★★★★★★★★★★★★★★★

Why are so many in the backyard looking for four leaf clovers, when opportunity, is knocking at the front door.

★★★★★★★★★★★★★★★★★

If your *WORK* seems dull that does not mean that you are in the wrong *WORK*. A dull heart will make any *WORK* seem dull. More likely you need a change of heart instead of a change of *WORK*.

. . . Scrapbook

★★★★★★★★★★★★★★★★★

Time is one thing that can not be retrieved. Use it or you'll loose it.

How to become a *MASTER?* Try patience, hard *WORK*, long hours, *PERSERVERANCE.* These are the price of *MASTERSHIP.* A master goes on long after others quit trying.

. . . Scrapbook

PEOPLE can be divided into three groups; those who make things happen, those who watch things happen and those who wondered what happened.

There is a way to get along without *WORK* in this world but the trouble is that while you are getting along without *WORK*, *YOU* are also getting along without almost everything else that is worthwhile.

TAKE TIME

Take time to live . . . That is what time is for . . . to live.
    Killing time is suicide.
Take time to *WORK* . . . it is the price of *SUCCESS.*
Take time to think . . . it is the source of power. Take time to play . . . it is the secret of youth.
Take time to read . . . it is the fountain of *WISDOM.*
Take time to be friendly . . . it is the road to *HAPPINESS.*
Take time to dream . . . it is hitching your wagon to a star.
Take time to love and be loved . . . it is the privilege of the Gods.
Take time to look around . . . it is too short a day to be selfish.
Take time to laugh . . . it is the music of the soul.
Take time to play with children . . . it is the joy of joys . . .

The Isle Service News . . . Scrapbook

WANTED: What this nation needs today is more men who can find things to be done without a manager and three assistants: men who listen when spoken to, who ask only enough questions to accurately carry out instructions, who move quickly without noise or fluster, who look you straight in the eye and tell the truth everytime, who get to *WORK* on time and never imperil the lives of others trying to be first out when day is done: who do not pity themselves because they have to *WORK*.

. . . Scrapbook

★★★★★★★★★★★★★★★★★

A man is not paid for his brains, but for using them.

★★★★★★★★★★★★★★★★★

WANTED: A man who is cheerful, courteous to every one and determined to make good. If interested, apply any hour, any place, to anyone.

. . . Motor West

★★★★★★★★★★★★★★★★★

At one Luther Burbank personally conducted 6,000 experiments. He raised 1,000,000 plants a year for his experimental purposes.

★★★★★★★★★★★★★★★★★

While employed as a mechanic at $150 dollars a month, Henry Ford spent evenings in a barn *WORKING* on a horseless carriage. It required eight years to develop a good motor.

Edison made thousands of experiments in developing inventions. Once he spent two years and two million dollars on an invention of little value. He developed the incandescent lamp in three days and three nights with no sleep.

My own thoughts pervade here: Did these three inventors have more *INITIATIVE,* or more *PERSEVERANCE,* or *"BUSHELS OF BOTH"* to make them *WORK* so arduously.

WORK thou for pleasure, paint, sing or carve.
Who *WORKS* for glory often misses the goal.
Who *WORKS FOR MONEY, COINS HIS VERY SOUL.*
WORK FOR WORKS SAKE, then and it may be,
That these things shall be added unto thee.

. . . Cox

Don't be like the salesman who said he honored all religions and all people—so he observed a day of rest on Monday with the Greeks, on Tuesday with the Persians, on Wednesday with the Egyptians, on Friday with the Turks, on Saturday with the Jews and on Sunday with the Christians.

. . . Scrapbook

★★★★★★★★★★★★★★★★★★

WORK has the mark of divine approbation since *HAPPINESS* is conferred not upon idleness but upon labor.

★★★★★★★★★★★★★★★★★★

Study to eradicate errors in your *WORK,* to know yourself and others. Feed your mind, it has clamored long enough. Ability breeds reliability, endurance and action.

> . . . Scrapbook

Our minds are like fountain pens—neither will *WORK* until you get something in them.

★★★★★★★★★★★★★★★★★★

Even if you're on the right track, you will get run over if you just stand there.

> . . . Arthur Godfrey

★★★★★★★★★★★★★★★★★★

Luck is only pluck, to try things over and over.
Courage and will, patience and skill
Are the *LUCK* of the four leaf clover.

> . . . Scrapbook

The fellow that takes little interest accomplishes but little, is paid but little and lasts but a little while.

★★★★★★★★★★★★★★★★★★

Age doesn't matter unless you're a cheese.

★★★★★★★★★★★★★★★★★★

THE BUSY MAN'S CREED . . . by Elbert Hubbard

I believe in the stuff I am handing out, in the firm I *WORK* for, and in my ability to get results.

I believe in *WORKING*, not weeping, in boasting, not knocking; and in the pleasure of my job.

I believe that a man gets what he goes after, that one deed done today is worth two deeds tomorrow, and that no man is down and out until he has lost faith in himself.

I believe in today and the *WORK* I am doing; in tomorrow and the *WORK* I hope to do, and in the sure reward which the future holds.

I believe in courtesy, in kindness, in generosity, in good-cheer, in friendship and in honest competition.

I believe there is something doing somewhere for every man ready to do it.

I BELIEVE I'M READY——RIGHT NOW!

★★★★★★★★★★★★★★★★★

The only *SUCCESSFUL* substitute for *WORK* is a miracle.

★★★★★★★★★★★★★★★★★

Labor disgraces no man, but many men disgrace labor.

★★★★★★★★★★★★★★★★★

IT IS A PLEASURE TO MEET PEOPLE——
Who do not talk about themselves,
Who acknowledge their mistakes,
Who are thoroughly dependable,
Who practice what they preach.

Who are habitually cheerful,
Who believe in the dawn of a better day,
Who are always sincere,
Who are quick to right a wrong.

Who avoid pretentiousness,
Who know what they are talking about,
Who are always on time,
Who are uniformly considerate.

Who fullfill their obligations,
Who are engaged in useful *WORK*,
Who are self disciplined,
WHO DO NOT DALLY AT THE DOOR.

.. Scrapbook

The favorite disguise of opportunity is hard work.

It is better to have one man *WORKING* with you than three men *WORKING* for you.

★★★★★★★★★★★★★★★★★

Blessed is he who has found his *WORK*.

... Carlyle

★★★★★★★★★★★★★★★★★

SIX LAWS OF WORK
1. A man must drive his energy, not be driven by it.
2. A man must be master of his hours and days.
3. The way to push things through must be learned.
4. A man must earnestly want.
5. Never permit failure to become a habit.
6. Learn to adjust to the conditions you have to endure but make a point of trying to alter all conditions to be favorable to you.

... Scrapbook

NINE WAYS TO RESPOND TO RESPONSIBILITY:
I won't is a tramp. I can't is a quitter. I don't is too lazy.
I wish I could is a wisher. I might is waking up. I will try is
on his feet. I can is on his way. I will is at *WORK—I DID
IS NOW THE BOSS.*

. . . Scrapbook

NO FEAR ON THE TRACK
Where are you going, what way do you ride?
  Are you facing the front or off to the side.
It isn't too late to change your direction;
  Your labor can be swayed by the boss of the section,
To the place or the rail that you are needed the most.
  Choose a spot that is weakened and set up your post.
Use your hands and your heart to help someone fight back.
  WORK well with the gang, and there'll be *NO FEAR ON
THE TRACK.*

. . . E. G. G.

★★★★★★★★★★★★★★★★★★

WORK: I AM the foundation of all business.
I AM the fount of all prosperity.
I AM the parent of genius.
I AM the salt that gives life its savor.
I HAVE laid the foundation of every fortune in America,
from the Rockefellers down.
I MUST be loved before I can bestow my greatest blessings
and achieve my greatest ends.
LOVED, I make life sweet, purposeful and fruitful.
I CAN do more to advance a youth than his parents, be
they ever so rich.
FOOLS hate me, wise men love me.
I AM represented in every loaf of bread that comes from
the oven, on every train that crosses the continent, in
every newspaper that comes from the press.

I AM the mother of *DEMOCRACY. ALL PROGRESS* springs from me. *WHO AM I? WHAT AM I? I AM WORK.*

. . . Scrapbook

★★★★★★★★★★★★★★★★★

When you go into the field to plow, be sure you take the Lord with you. When you go about your business, be sure the Lord is with you, also in your dealings with men.

Brigham Young to the Mormon Saints

★★★★★★★★★★★★★★★★★

WORK AND WIN

I've seen a lot o' fellows try a lot of different ways for carvin'
   out their fortunes, through my little stretch of days;
I've watched the clever-minded an' the idle gossiper and I've
   noticed at the finish after all the fuss and boast,
That the chap who *WORKS* the hardest is the one who gets
   the most!
There ain't no way round it—it's the man who never stops,
   But keeps right on a-farmin' that will have the biggest
   crops.
Oh, there's times it's hot for hoein', an' there's days the fish
   'll bite
But the field that's been neglected never does come through all
   right;
An' what is true of farmin' must be true of ever post,
It's the chap who *WORKS* the hardest that *ALWAYS GETS
THE MOST.*

. . . *Edgar A. Guest*

★★★★★★★★★★★★★★★★★

Greatness lies not in being strong but in the right use of strength.

No race can prosper till it learns that there is as much dignity in tilling a field as in writing a poem.

★★★★★★★★★★★★★★★★★

YOUR JOB

Whenever you're working—in office or shop,
  And however far you may be from the top—
And though you may think you're just treading the mill,
  Don't ever belittle the job that you fill;
For, however little your job may appear,
  You're just as important as some little gear
That meshes with others in some big machine,
  That helps keep it going— though never is seen.
They could do without you—we'll have to admit—
  'But business keeps on, when the big fellows quit!
And always remember, my lad, if you can,
  The job's more important—oh yes— than the man!
So if it's your hope to stay off the shelf,
  Think more of your job than you do of yourself,
Your job is important—don't think it is not—
  So try hard to give it the best that you've got!
And don't think ever you're part of little account—
  Remember, you're part of the total amount.
If they didn't need you, You wouldn't be there—
  So always, my lad, keep your chin in the air.
A digger of ditches, mechanic or clerk—
*THINK WELL OF YOUR COMPANY, YOURSELF AND
  YOUR WORK!*

. . .Scrapbook

★★★★★★★★★★★★★★★★★

FACE THE SUN

Don't hunt after trouble, but look for *SUCCESS.*
  You'll find what you look for, don't look for distress.
If you see but your shadow, remember I pray,
  That the sun is still shining, but you're in the way.

Don't grumble, don't bluster, don't dream and don't shirk,
  Don't think of your worries, but think of your *WORK*.
The worries will vanish, the *WORK* will be done,
  No man sees his shadow, who *FACES THE SUN*.

. . . Scrapbook

★★★★★★★★★★★★★★★★★

Thank God every morning when you get up that you have something to do today which must be done whether you like it or not. Being forced to *WORK* and do your best will breed in you *TEMPERANCE, SELF CONTROL, DILIGENCE, STRENGTH OF WILL, CHEERFULNESS, CONTENT* and a hundred virtues which the idle never know.

. . . Rob Kingsley

★★★★★★★★★★★★★★★★★

WORK is the true elixir of life. The busiest man is the happiest man. Excellence in any art of profession is attained only by hard *WORK*. Never believe that you are perfect. When a man imagines, even after years of striving that he has attained perfection, his decline begins.

★★★★★★★★★★★★★★★★★

If you look for new challenges, your *WORK* can become a chain reaction. Each link, connected with *WORK*, will secure man's link of *HOPE* to *REALITY*.

. . . E. G. G.

★★★★★★★★★★★★★★★★★

I remember an article, a treasure of a thought, that was written by Ann Landers. She advised, "Teach your children there is dignity in hard *WORK* no matter how performed, by calloused hands or skilled fingers. Teach them that a useful life is blessed and a pleasure seeking life is empty." I can not recall it entirely, but I know that every parent should have this advise in their book of rules for bringing up a precious child.

What you are and what you have, represents a veritable *GOLD-MINE*—provided that you exercise your talents a sufficient number of times each day— in the proper places, at the proper time. If you take your talents out of your salesman's bag one at a time, it will take forever to complete a days *WORK. THINK BIG, TALK BIG AND WORK BIG. WHAT YOU DO WITH YOUR PRESENT OPPORTUNITIES, ABILITIES AND ASPIRATIONS, YOU* will likely do with larger abilities and desires in the future. *DO NOT MISLEAD* yourself into believing that under different circumstances or in another environment, you could attain more and achieve better.

Anyone who believes that he was not sufficiently endowed, by God, with certain capabilities and faculties should take a second look. *IT IS YOUR RESPONSIBILITY TO:* utilize and manage your potentiality, to "open your eyes to opportunity," to "stand on your own two feet," to project your own goals, to eliminate fear and worry with *WORK* and to make *YOURSELF* a respectable and honorable individual.

There is a song I learned as a child that always activated my motives. It is *"WORK FOR THE NIGHT IS COMING, WORK FOR THE COMING DAY."* Don't put off your labors, whether they are unpleasant or enjoyable. Put "your shoulder to the wheel" and "your hand to the shovel." Whatever mountain or mine you need to climb or dig in, *NOW IS THE TIME* to start pushing and digging. *RIGHT NOW, WHERE EVER YOU ARE-NOW IS THE PLACE TO DO YOUR BEST WORK.* "Action is the parent of results, dormancy is the brooding Mother of discontent." Any dependable person who pushed himself through the

TUNNEL OF WORK
BECOMES
THE VOICE
OF EXPERIENCE.

★★★★★★★★★★★★★★★★

Experience

PART III

EXPERIENCE

A general opinion often observed is that *EXPERIENCE* comes with age. *EXPERIENCE* is knowledge gained by trial and practice through *WORK*, physical and intellectual.

★★★★★★★★★★★★★★★★★★

Few things are impossible to diligence, skill and *EXPERIENCE.* Any good *SALESMAN*, any good miner knows this. Diligence and *EXPERIENCE* are the Mother of good fortune. What your future and fortune has in store for you depends on what you place in store for the future.

★★★★★★★★★★★★★★★★★★

Take the tool of *EXPERIENCE* firmly in your hand and *DIG* with me. It is wisest and sometimes the hardest teacher. But harken! lest you fail to consider how much it could cost you. *EXPERIENCE* is the dear teacher that molds you, young or old.

EXPERIENCE is the best teacher, but the tuition is much too high.

EXPERIENCE is a good school, but it never gives you a vacation.

EXPERIENCE is the only teacher that gives the test first and the lesson later.

There are no free scholarships in the school of *EXPERIENCE*.

EXPERIENCE is one of the few things in life that you can't get on the easy payment plan.

EXPERIENCE can seldom be bought on the bargain counter.

EXPERIENCE comes with age, which is the time it does you the least good.

EXPERIENCE shows that *SUCCESS* is due less to ability than to zeal. The winner is he who gives himself to his *WORK*, body and soul.

EXPERIENCE is the best method of acquiring knowledge because everyone gets individual instruction.

The trouble with *EXPERIENCE* is that few people are born with it and you never have it until after you need it.

★★★★★★★★★★★★★★★★★

RULES FOR LIVING   *. . .Grenville Kleiser*

PLAY THE GAME, know the rules and observe them.
ALWAYS PLAY FAIR, no matter what the other man may do.
PLAY TO WIN. If you lose, be a good loser. But don't lose.
Study life as you would a game of chess. Know the moves, the dangers, difficulties and the choicest rewards. Remember that no allowance is made for ignorance. If you find you don't know the moves and the rules for life's game, you must learn through *DISCIPLINE* and *EXPERIENCE. YOU MUST PAY FOR MISTAKES.*

★★★★★★★★★★★★★★★★★

EXPERIENCE tells you not to repeat an offense, lest greater punishment befall you. *EXPERIENCE* is the greater *SCHOOLMASTER OF LIFE*, constantly disciplining you through tests, trials and difficulties. It imposes penalties for mistakes and misjudgements. It indicates where danger lies and warns you of inevitable punishment lurking in undesirable directions. Thus you learn the penalty for waste of time, the scattering of energies, the pursuit of illconsidered plans, and the prodigal squandering of physical and mental power. *EXPERIENCE* makes you understand that you are living in a real world with real *WORK* to do.

. . . Scrapbook

The words you have read on the *SCHOOLMASTER* of life are from Dad's book and has no author listed. However, I feel that he must be a person who has met challenges and hard *WORK*; who has searched and dug in many mines of *EXPERIENCE*. His tools must be many, his rewards must be *HAPPINESS* and *SUCCESS*. He, or she, must be a *CHARACTER* I would like to know.

With a *WISE* choice of classrooms, with the *SCHOOLMASTER OF EXPERIENCE*, with daily *WORK* and study, an ambitious student of any age, can graduate with a diploma of good *CHARACTER*.

SUCCESS

Wisdom

Perseverance

Initiative

Enthusiasm

Personality

A Man's Fortunes
Are The
Fruits
Of His
CHARACTER

Emerson

PART IV

CHARACTER LADDER

CHARACTER is a series of traits which constitute a Ladder of moral excellence. *WORK* and *EXPERIENCE* are the prime contractors in its construction. It is braced by five sturdy rungs, *PERSONALITY*, *ENTHUSIASM*, *INITIATIVE*, *PERSEVERANCE* and *WISDOM*. It has been said, "the bottom rung of a ladder should be its strongest, it supports the most people." More people step on the first but do not make it to the next. Should you be *WISE* enough to, you will have in your possession five of the most valuable and profitable tools with which you can *"DIG YOUR OWN GOLD."*

★★★★★★★★★★★★★★★★★

CHARACTER is what you are in the dark.

★★★★★★★★★★★★★★★★★

Our *CHARACTER* is our will, for what we will, we are.

★★★★★★★★★★★★★★★★★

The man who holds the *LADDER* firmly at the bottom is about as important as the man at the top.

★★★★★★★★★★★★★★★★★

Life's *LADDER* is full of splinters. They hurt most when coming down.

A LADDER is anything by which one climbs or ascends. I challenge you to read these five steps on my *LADDER* and then *"CLIMB, FAR, YOUR GOAL THE SKY, YOUR AIM THE STARS."*

★★★★★★★★★★★★★★★★★★

CHARACTER   STEP I _ _ _ PERSONALITY
PERSONALITY is a collection of individual traits, your most unique possession.
> . . . Mahatmi Ghandi

★★★★★★★★★★★★★★★★★★

A person completely wrapped up in himself makes a very small package.
> . . .H. E. Fosdick

★★★★★★★★★★★★★★★★★★

When wealth is lost, nothing is lost, when health is lost, something is lost, when *CHARACTER* is lost, all is lost.

★★★★★★★★★★★★★★★★★★

Wealth can not create *PERSONALITY,* but *PERSONALITY* can create wealth. *PERSONALITY* can not be bought, inherited, delegated or borrowed.

★★★★★★★★★★★★★★★★★★

The greatest pleasure is to do a good action by stealth and have it found out by accident.

It is not enough to possess virtue,
it should be practiced.

★★★★★★★★★★★★★★★★★

Hang on to your temper, no one else wants it.

. . .Jay Ackerman

★★★★★★★★★★★★★★★★★

Temper is what gets most of us into trouble, pride is what
keeps up there.

★★★★★★★★★★★★★★★★★

Men are like steel, when they lose their temper, they're
worthless.

★★★★★★★★★★★★★★★★★

Of all the things you wear,
your expression is most important.

★★★★★★★★★★★★★★★★★

Hens that cackle loudest, often lay the smallest eggs.

★★★★★★★★★★★★★★★★★

I would rather be a *COULD BE*, If I could not be an *ARE*,
  For a *COULD BE* is a *MAYBE*, with a chance of touching
    par.
I would rather be a *HAS BEEN*, than a *MIGHT HAVE BEEN*
  by far,
For a *MIGHT HAVE BEEN* has *NEVER BEEN*, while a *HAS*
  was once and *ARE*.

. Scrapbook

It isn't easy: to apologize, to face a sneer, to begin again, to forgive and forget, to admit error, to keep out of a rut, to be unselfish, but it always pays in the end.

. . .J. P. Fleishman

Honesty is the one card in the pack you can play at anytime without thinking of how to play it.

Habits are first like cobwebs, then a cable.

Laugh a little now and then, it brightens life a lot,
  You can see the brighter side, just as well as not.
Don't go mournfully around, gloomy and forlorn,
  Try to make your fellow man, glad that you were born.

★★★★★★★★★★★★★★★★★★

Remember: it is better to say something good about a bad man than to say something bad about a good man.

★★★★★★★★★★★★★★★★★★

He who is self-centered travels in very small circles.

T. Kirkwood Collins

★★★★★★★★★★★★★★★★★★

What is the golden rule for getting along with people? Simply this: see that there ego is not deflated.

I KNOW SOMETHING GOOD ABOUT YOU

Wouldn't this old world be better, if folks we met would say,
"I know something good about you and treat us just that way.
Wouldn't life be lots more happy, if we praised the good we
   see,
For there's such a lot of goodness, in the worst of you and me.
Wouldn't it be nice to practice, that fine way of thinking to?
You know something good about me? I know something good
   about you!"

★★★★★★★★★★★★★★★★★

Please, charm, delight and fascinate those you meet. Make
them feel that you enjoy their company. Let them talk. Listen
and compliment them when you can do so sincerely. Make
them feel important. However humble or famous they may be,
let them see your *EYES*, your *SMILE*, the expression of your
face or your bearing.
One who can do this in the right way has what is called a
charming *PERSONALITY*. Some people are born with the gift;
others must cultivate it. "The most important thing you may
wear is your *PERSONALITY*."

★★★★★★★★★★★★★★★★★

Many people are indifferent whether they please people or
not. This self-reliance is a handicap to progress. We get ahead
by inducing others to do things our way and to see things from
our point of view. If we are backed by law and money we can
force them to do as we wish, whether they like it or not, but
that method is expensive and burdensome. But, with percep-
tion and tact we can get our method over with but a fraction of
the effort.

★★★★★★★★★★★★★★★★★

Therefore, the man who is interested in practical *SUCCESS* will do his utmost to master the art of pleasing others. He will find that this art is the supreme labor-saver in our complicated civilization which makes us so dependent on one another.

. . . Scrapbook

★★★★★★★★★★★★★★★★★

Your *PERSONALITY* should bring a very fancy price, Backed by ability and brain to take and give advise;
But to do this you'll have to strive to do things right, not wrong—
Then wrap your products in a smile and sell them with a song.

. . . Scrapbook

★★★★★★★★★★★★★★★★★

Youngsters and adults sometimes have heads like door knobs. . .anybody can turn them.

★★★★★★★★★★★★★★★★★

The exercise that does you the least good is patting youself on the back.

★★★★★★★★★★★★★★★★★

He that falls in love with himself will have no rivals.

★★★★★★★★★★★★★★★★★

The bigger a man's head gets, the easier it is to fill his shoes.

★★★★★★★★★★★★★★★★★

Behave toward everyone as if receiving a great guest.

★★★★★★★★★★★★★★★★★

If you can laugh at yourself there's still hope that with the Lord's help you may still amount to something.

Trouble knocked at the door, but hearing a laugh within, hurried away.

★★★★★★★★★★★★★★★★★★

I used to tell my troubles to everyone I knew. And the more I told my troubles, the more my troubles grew.

. . . Scrapbook

★★★★★★★★★★★★★★★★★★

One thing to say to chronic sufferers of anxiety *"DROP DREAD."*

★★★★★★★★★★★★★★★★★★

WHEN EVENING COMES, go off into a quiet place and review your day.
HAVE YOU been kind and thoughtful or mean and thoughtless?
HAVE YOU kept an even temper or have you lost your temper when things have gone wrong?
HAVE YOU been pleasant or grouchy?
HAVE YOU inspired those whom you have met or have you depressed and discouraged them?
HAVE YOU done something creative and worthwhile or have you wasted the day with petty things?
HAVE YOU spent a few hours in study?
HAVE YOU increased the happiness-moments in the lives of others or have you thought only of yourself?
HAVE YOU enlarged your mental horizon, expanded your *PERSONALITY*—have you grown larger or shrunk smaller?
WHAT WE do day by day determines what we become.
HOUR BY HOUR we will build our lives for better or for worse. . . . Scrapbook

★★★★★★★★★★★★★★★★★★

Always do right. This will gratify some people and astonish the rest.

. . . Mark Twain

We seldom develop "I" trouble while looking for our own faults and other people's good points.

★★★★★★★★★★★★★★★★★★

BE FRIENDS with everybody. When you have friends you will know there is somebody who will stand by you. You know the old saying, that if you have a single enemy you will find him everywhere. It doesn't pay to make enemies. Lead the life that will make you kind and friendly to everyone you meet, and you will be surprised what a happy life you will lead.

. . . Charles Schwab . . . Scrapbook

★★★★★★★★★★★★★★★★★★

A smile increases your *FACE VALUE.*

★★★★★★★★★★★★★★★★★★

When you meet a person without a smile, give him one of yours.

★★★★★★★★★★★★★★★★★★

Your mind is a sacred enclosure into which nothing harmful can enter except by your permission.

. . . Arnold Bennet

★★★★★★★★★★★★★★★★★★

DEFINITION OF A GENTLEMAN: a kindly heart, a quiet voice, polite words and manners, a hand ready to help. Attention to the little things for the comfort of others. Freedom from anger, boasting and patronizing. Towards the strong—courage, towards the weak—chivalry, towards all men—fairness.

. . . Scrapbook

★★★★★★★★★★★★★★★★★★

DO YOU SEE ONE IN THE MIRROR? _ _ _ LOOK AGAIN!

★★★★★★★★★★★★★★★★★★

BE A BOOSTER, never a *KNOCKER*—and you will reap dividends in friendship and sales.

. . . R.K. Dugdale

★★★★★★★★★★★★★★★★★

No man is free who is not master of himself.

★★★★★★★★★★★★★★★★★

Before you step up to the next rung of the *LADDER*, contemplate on *PERSONALITY;* that which constitutes distinction of a person. Often we hear someone say, "She has a wonderful *PERSONALITY."* So what is this *PERSONALITY?* How do you get it? Can you buy it? Can you learn it out of a book? Is it both desirable and beneficial?

You can't buy it! No one can teach it to you! Watch the person with the "wonderful *PERSONALITY."* It has a simple formula. The secret is: he likes people. Learn to enjoy people, applaud their *SUCCESS*, share their misfortunes. Be sincere, if you can't feel and show real sincerity, *KEEP YOUR MOUTH SHUT.* Don't be the kind of a person who says, "If you like me, I'll like you." Learn to love and appreciate people. By most individuals, you are judged or criticized by the amount of friends you have. To have a fanfare of true friends, you will beyond a doubt, have also, a *DELIGHTFUL AND PLEASANT PERSONALITY.*

If your *PERSONALITY, WITH* all its attractions and attributes, is lying dormant, get it out and dust it off. Everyone has the power to row his own boat, provided he uses the oars of *VIGOR* and *VITALITY.* With each stroke rises your *OPPORTUNITY* to be the kind of person that your co-workers and friends will label *"BUBBLING OVER WITH PERSONALITY AND ENTHUSIASM. YOU ARE CAPABLE OF BEING THE CAPTAIN AT THE HELM OF A HAPPY SHIP.*

★★★★★★★★★★★★★★★★★

Enthusiasm

CHARACTER STEP II

ENTHUSIASM

Nothing great was ever achieved without *ENTHUSIASM*.

★★★★★★★★★★★★★★★

Like the chicken and the egg,
ENTHUSIASM and *SUCCESS* go together.

★★★★★★★★★★★★★★★★

You can not kindle a fire in another heart until it is burning
in your own.

★★★★★★★★★★★★★★★★

WIN-LOSE-OR-DRAW

A SALESMAN (that's you) whatever he sells; ability,
PERSONALITY or commonsense, finds what's really in him
when he runs into tough traveling that tempts him to lie down
and quit. Those are the times he learns whether he is kidding
himself by claiming he is a *SALESMAN*. Right there he gets
the kick out of his *WORK* and that's where the real
SALESMAN begins to *GRIN AND WADE IN!* If he has the
right stuff in him, such as *ENTHUSIASM, INITIATIVE* and
SELF-CONFIDENCE he will have a lot of fun, gain friends and
much *EXPERIENCE*, no matter what the results may be -
WIN-LOSE-OR-DRAW.

. . . Scrapbook

★★★★★★★★★★★★★★★★

Lugging around a grievance wears out your *ENTHUSIASM*,
slows up your *WORK* and spoils your day. Unload and go on.

Are you in earnest? Seize this very minute. What you do or dream you can, *BEGIN* it and the *WORK WILL BE COMPLETED.*

An artist was asked which was the greatest of all his productions. His reply, "The next one." No painter, musician or writer would ever finish a production if they did not have the *ENTHUSIASM* installed in the brush or pen to begin again. You don't have to produce a work of art to have it. Even the little ant has it. There-in is a lesson in *ENTHUSIASM* and *PERSEVERANCE* which fascinates me.

A captain of industry once said he wanted everyone of his organizations to be fired with *ENTHUSIASM*. If he had an employee not fired with *ENTHUSIASM*, he would try to fire him with *ENTHUSIASM*, but if he could not fire him with *ENTHUSIASM*, then, indeed he would then fire him with *ENTHUSIASM*.

A tough job is only tough in proportion to the way you look at it. If you can't see anything but the disagreeable angle to a difficult task, it will be hopeless. You are simply betting yourself you cannot do it—and then of course, you can't. *DIG OUR YOUR ENTHUSIASM*. "A winner never quits, a quitter never wins."

Few things are as bad as *ENTHUSIASTIC* ignorance.

ENTHUSIASM is the genius of sincerity and the truth accomplishes no victories with it.

★★★★★★★★★★★★★★★★★★

PSALM FROM A SALESMAN'S BIBLE

AND IN THOSE DAYS, behold there came through the gates of the city, a salesman from afar off, and it came to pass, as the day went by, he sold plenty. They that were grouches smiled on him and gave him the hand that was glad. The tightwads opened their purses to him.

AND IN THAT CITY were they that were the order takers and they that spendeth their days in adding to the alibi sheet. Mightily were they astonished. They said one to the other, "What the heck; how doth he get away with it?" And it came to pass that many gathered in the back office and a sooth sayer came among them. And he was one wise guy. And they spoke and questioned him saying, "How is it that this stranger accomplisheth the impossible?"

WHEREUPON the sooth sayer made answer; "He of whom you speak is the one hustler. He ariseth very early in the morning and goeth forth full of pep. He complaineth not, neither doth he slouch. He is arrayed in purple and fine linen, while ye go forth dirty and faces unshaven. . .

WHILE YE gather here and say one to the other, "Verily! this is a terrible day's *WORK,*" he is already abroad. And when the eleventh hour cometh, he needeth no *ALIBIS.* He said not, to the mass, "Behold they that are in this town are a bunch of boneheads, nor doth he report that they can not be sold to."

"HE KNOWETH his line and they that would stall him off, they buy from him. Men say unto him, "Nay, nay" when he cometh in, yet when he goeth forth he hath their names on the line that is dotted."

"HE TAKETH with him three angels—*ASPIRATION, ENTHUSIASM AND PERSPIRATION.* He knoweth whereof he speaketh and he *WORKETH* to beat all. Verily, I say unto you, *GO AND DO LIKEWISE.*

. . . Scrapbook

★★★★★★★★★★★★★★★★★★

There is something *GLORIOUS* in a job well done—an exhilarating feeling from the soul, that is above money, wares and men. Perhaps you are experiencing an awakening. You might think it a little late, but not too late, to make up for some of the time you have wasted. Get under way. Weak resolutions never materialize into anything worthwhile. The very fact that you have come to a realization of how you have failed to put your job over to the best of your ability, is the first point necessary for your own improvement.

. . . Scrapbook

★★★★★★★★★★★★★★★★★★

THE WHOLE SECRET OF REMAINING YOUNG in spite of years and even of gray hairs, is to cherish *ENTHUSIASM* in oneself, by poetry, by contemplation, by charity, by the maintenance of *HARMONY IN THE SOUL. WORKING* with young people in *"MY DIGGING OPERATION"* gives me spurts of energy and *ENTHUSIASM.* I have to stay young to keep up with them. *TRY IT! IT WORKS!*

★★★★★★★★★★★★★★★★★★

YOU ARE THE YEAST CAKE

ENTHUSIASM in our association is like yeast cake in the
   dough;
Each depends upon the other, if we want to make things go.
Dough would always just be heavy, if you left it to itself,
And the yeast cake too, is useless. when left upon the pantry
   shelf.

When you mix the two together, *ZIP!* the stuff begins to grow.
There is life and zeal about it, soon its *WORK* begins to show.
YOU ARE THE YEAST CAKE, YOUR COMPANY THE DOUGH.
We must depend upon each other, if we want to make things go.
So let's mix the two together; *ZIP!* We will make our association grow.

> . . . Alsada Potts . . . Scrapbook

Lightning strikes with *ENTHUSIASM* and usually hits its mark. This does not justify using your *ENTHUSIASM* with such impetuous force. Better to let your *ENTHUSIASM* radiate with the glow of the rainbow.

> . . . E. G. G.

★★★★★★★★★★★★★★★★★

Raise up, don't wait until your yeast cake has lost it's *ZIP*. Take it off the shelf and use your *INITIATIVE*, which is the power of commencing, to start your *DOUGH*. "Even a journey of a 1,000 miles starts with a single step." Put your foot down with *ENTHUSIASM* and finish with *INITIATIVE*.

> . . . *E. E. G.*

★★★★★★★★★★★★★★★★★

Initiative

CHARACTER STEP III

INITIATIVE

INITIATIVE is doing the right thing without being told.

★★★★★★★★★★★★★★★★★★

It is better to light one small candle, than to curse the darkness.

. . . Confucius

★★★★★★★★★★★★★★★★★

People are always blaming circumstances. The people who get on in the world are people who get up and look for the circumstances they want and if they can't find them, make them.

. . . George Bernard Shaw

★★★★★★★★★★★★★★★★★★

The most underdeveloped territory in America
is under men's hats.

. . . Norman Dryden

★★★★★★★★★★★★★★★★★★

There is always room at the top, but the elevator is not always running. You must walk up the street on your own two feet.

. . . David Starr Jordan

★★★★★★★★★★★★★★★★★★

The great thing in this world is not so much where we stand but which direction we're moving.

. . . O.W. Holmes

It's true you can't take it with you, but folks ought to remember that where you got it may determine where you go.
. . . Richard L. Evans

★★★★★★★★★★★★★★★★★★

It is your duty to make the most of the best thats in you.
★★★★★★★★★★★★★★★★★★

If you have no ambition,
excuse me for breaking into your slumber.

★★★★★★★★★★★★★★★★★★
IT'S ALL UP TO YOU
You are the fellow who has to decide
Whether you'll do it or toss it aside;
You are the fellow who makes up your mind
Whether you'll lead or will linger behind,
Or be contented to stay where you are.
Take it or leave it. There's something to do!
Just think it over. *IT'S ALL UP TO YOU.*

. . . Scrapbook

★★★★★★★★★★★★★★★★★★

No working parts are missing from the self made man.

★★★★★★★★★★★★★★★★★★

LUCK and *INITIATIVE* make good fishermen! Do you need help to row your boat?

★★★★★★★★★★★★★★★★★★

INITIATIVE was responsible for three of the greatest accomplishments of the world:
1. Thomas A. Edison . . . Wanted: A man to light the world.
   He did.

2. Charles Lindbergh . . . Wanted: A man to fly the Pacific. He did.
3. Abraham Lincoln . . . Wanted: A man to free the slaves. He did.

★★★★★★★★★★★★★★★★★★

You can't get bread out of an empty cupboard, nor ideas out of an empty head.

★★★★★★★★★★★★★★★★★★

All things are apt to come to him who doesn't wait.

★★★★★★★★★★★★★★★★★★

LUCK!!!

Oh, there isn't a doubt that the fellow who wins
    May need the assistance of luck,
When the going is hard and the struggle begins
    To help out his strength and his pluck
Every observer of life, here must own
    That chance sometimes raises men high,
But the thing men call LUCK has never been known
    To help out the chap who won't try.
The man who goes fishing when all signs are wrong
    To catch a fish, he lucky may be,
But the idler refusing to follow along
    And take his boat and row out to sea,
Won't have any fish when the night settles down
    To cook, eigher large ones or small;
For the idler who chooses to stay in town
    Good *LUCK* can do nothing at all.
That some men are lucky there's none can deny,
    But of them it usually is true
They're the fellows full-willing to get out and try
    What many a scoffer won't do.
They don't sit and wail that signs are all wrong,
    They get out and hustle and *WORK* and they fight.
THEY'RE ON HAND WHEN LUCK HAPPENS ALONG.
. . . Scrapbook

DEPEND UPON YOURSELF. Make your judgements trustworthy. You can develop judgement as you do the muscles of your body by judicious daily exercise.

                                    . . . Grenville Kleiser

In todays business there is too little of the idea of personal responsibility and too much of the idea that the world owes us a living, forgetting that if the world owes us a living we must be collectors . . . *AND GOOD ONES.*

What I need is someone to get me to do the things I ought. Only one person can do this. It has to be an *"INSIDE JOB."*

                                    . . . Emerson

There is no gathering the rose without being pricked by the thorns . . . (Take hold and try again.)

It is a striking coincidence that the word *AMERICAN* ends in I Can.

The itching sensation that some people mistake for ambition is merely inflammation of the wishbone.

Don't be contented with your lot, until it is a lot more.

★★★★★★★★★★★★★★★★★

Not everything that is faced can be changed, but nothing can be changed until it is faced.

. . . James Baldwin

★★★★★★★★★★★★★★★★★★

If a man does not know to what port he is steering, no wind is favorable to him.

★★★★★★★★★★★★★★★★★★

Nothing ever built arose to touch the skies unless some man dreamed that it should—some man believed that it could and some man willed that it must.

. . . Charles F. Kettering

★★★★★★★★★★★★★★★★★

You can not run away from a weakness; you must sometime fight it or perish, and if that be so, *WHY NOT NOW?*

★★★★★★★★★★★★★★★★★

The people who tick but never tell time are part of the waste material of the world.

★★★★★★★★★★★★★★★★★

It isn't so much the size of the dog in the fight as the size of the fight in the dog. How's your *FIGHT?*

WISHBONE is nice to have,
but don't use it in place of *BACKBONE*.

One professor described a student as having the greatest of gifts but was too lazy to unwrap them.

No age or time of life, no position or circumstances has a monopoly on *SUCCESS*. Any age is the time to start doing something.

... Wynn Johnson

★★★★★★★★★★★★★★★★★★

Anyone who lives at less than his best, cheats himself. Make the most of yourself, for that is all there is to you.

★★★★★★★★★★★★★★★★★★

For most of us, there is little chance of stumbling on to a rich mine. Loosen your back-pack and get out the "I" tools, *INITIATIVE*, *INSPIRATION*, and *IMPLEMENTATION*. You can't dig gold without *BACKBONE*, *A HOPEFUL HEART* and *HARD WORK*.

★★★★★★★★★★★★★★★★★★

A STITCH IN TIME saves nothing
if your needle isn't threaded.

... John D'Ort

★★★★★★★★★★★★★★★★★★

Life is not just a bowl of cherries, you have to climb the tree and bark your shins to get the most out of the bowl.

★★★★★★★★★★★★★★★★★★

Competitors pay dividends out of business that you overlook.

★★★★★★★★★★★★★★★★★★

Delay is responsible for more failures, more lack of *SUCCESS* than any other circumstance.

★★★★★★★★★★★★★★★★★★

Few of us get dizzy from doing too many good turns.

★★★★★★★★★★★★★★★★★★

Opportunities are never lost. The other fellow takes those you miss.

★★★★★★★★★★★★★★★★★★

I have found that *INITIATIVE* is like arithmetic. If you add it to knowledge, experience, skill and *WISDOM* ,you have the sum of *SUCCESS*. Should you subtract it from these, you will find that you will zero in on *FAILURE*. If you multiply these facets with *INITIATIVE,* you will get a grand total of *HAPPINESS AND SUCCESS. INITIATIVE* is the steam of your locomotive. The more steam generated, the more rapidly the train gets to its destination. Your life's train won't be "comin down the track" unless you get up steam. How much you generate will be determined by your *PERSEVERANCE.*

Perseverance

CHARACTER STEP IV

PERSEVERANCE

Steadfastly one must climb the mountain with
PERSEVERANCE. *INITIATIVE* gives *PERSEVERANCE* an
extra shove of energy.

★★★★★★★★★★★★★★★★★

If you only knock long enough and loud enough at the gate,
you are sure to wake somebody up.
.˙. Henry Wadsworth Longfellow

★★★★★★★★★★★★★★★★★

Great works are performed not by strength but by
PERSEVERANCE.

★★★★★★★★★★★★★★★★★

There was a small boy learning to skate who kept falling
down. A bystander said, "Sonny, you're getting all banged up,
why don't you sit down and watch?" The little boy
answered, "Mister, I didn't get these skates to give up on. I got
them to learn on.

★★★★★★★★★★★★★★★★★

Wake up, get up, stay up, never give up, let up or back up.

I took a job without asking, I took a job and stuck. I took a chance they wouldn't, Now they call it *LUCK*.

★★★★★★★★★★★★★★★★★

FAILURE is the path of least resistance.
★★★★★★★★★★★★★★★★★

When you trip, fall forward and get farther along.

★★★★★★★★★★★★★★★★★

No man knows his real possibilities, the truly great things he can accomplish until he makes up his mind to go at it in earnest.

...J. Wightman

★★★★★★★★★★★★★★★★★

Many ball games have been won in the last inning.

★★★★★★★★★★★★★★★★★

Leadership is won by ordinary men with more than ordinary determination who keep everlastingly at it.

★★★★★★★★★★★★★★★★★

Thomas Edison advised that he tried over 10,000 experiments. What if he had quit at fifty or one hundred?

★★★★★★★★★★★★★★★★★

Nothing in the world can take the place of *PERSISTANCE.*
Talent will not. *PERSISTANCE* and determination alone are
omnipotent. The watchword *"PRESS ON"* has and will solve
the problems of the human race.

. . . Calvin Coolidge

Lure yourself into the channel of *SUCCESS.* Endurance,
patience and inclination are necessary to swim the channel.
Wishing and longing for victory won't acquire it. Overhand
strokes of zeal, insatiable desire and ambition will pace the
final accomplishment. We don't win by swimming underhand.
Having a goal and sticking to it, like the "postage stamp,"
WILL GET YOU THERE. "Consider the postage stamp, its
usefullness is its ability to stick to one thing until it gets
there." This motto I like.

Keep it in mind that you are going to be something more
than an average man, that you are going to get your head at
least a little way above the heads of those around you.

The real difference between men is energy. A strong will, a
settled purpose and invincible determination can accomplish
almost anything; in this lies the distinction between great men
and little men.

YOU CAN DO as much as you think you can,
    but you'll *NEVER ACCOMPLISH MORE;*
If you're afraid of yourself, young man,
    There's little for you in store,

For *FAILURE* comes from the inside, first,
  It's there if we only knew it.
And you can win, though you face the worst,
  If you *FEEL* that you're going to do it.

                                    . . . Scrapbook

★★★★★★★★★★★★★★★★★

A man who is satisfied with potluck usually gets just that.

★★★★★★★★★★★★★★★★★

Not everyone who climbs reaches the top, but the man who keeps climbing, keeps getting nearer. *"PERSIST, TO THE TOP OF THE CHARACTER LADDER.*

★★★★★★★★★★★★★★★★★

When Noah sailed the waters blue,
  He had his troubles same as you.
For forty days he sailed his ark,
  Before he found a place to park.

                                    . . . Scrapbook

★★★★★★★★★★★★★★★★★

Do more than exist—*LIVE.*
Do more than look——*OBSERVE.*
Do more than read——*ABSORB.*
Do more than hear——*LISTEN.*
Do more than listen——*UNDERSTAND.*
Do more than think——*PONDER.*
DO MORE THAN TALK——SAY SOMETHING.

                                    . . . Scrapbook

★★★★★★★★★★★★★★★★★

AND WHEN YOU DO, SPEAK WORDS OF WISDOM.

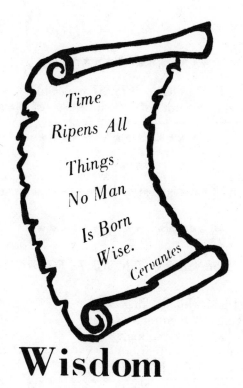

Time
Ripens All
Things
No Man
Is Born
Wise.
Cervantes

Wisdom

CHARACTER STEP V

WISDOM

A WISE man is like a straight pin, his head keeps him from going too far.

. . . Scrapbook

★★★★★★★★★★★★★★★★★

It is probably *WISEST* never to
insult an alligator until after you've crossed the river.

★★★★★★★★★★★★★★★★★

It is a thousand times better to have common sense without education, than to have education without common sense.

★★★★★★★★★★★★★★★★★

HORSE SENSE from THOUGHTS FOR TODAY

A horse can't pull while kicking,
    This fact I merely mention,
And he can't kick while pulling,
    Which is my chief contention.
Let's imitate the good old horse,
    And live a life that's fitting.
Just pull an honest load and then,
    There'll be no time for kicking.

. . . Scrapbook

The height of human *WISDOM* is to bring our tempers down to our circumstances and to make a calm within, under the weight of the greatest storm without.

... Daniel Defoe

★★★★★★★★★★★★★★★★★

If any of you lack *WISDOM*, let him ask of God, that giveth to all men liberally, and upbraideth not; and it shall be given him.

... James 1:5

★★★★★★★★★★★★★★★★★

We learn *WISDOM* from *FAILURE*
much more than from *SUCCESS.*

★★★★★★★★★★★★★★★★★

The *WISE* man becomes full of good even though he gathers it little by little.

★★★★★★★★★★★★★★★★★★

A thought once awakened,
never slumbers.

... Carlyle

★★★★★★★★★★★★★★★★★

A man should never be ashamed to say he has been wrong, which is really saying he is *WISER* today than yesterday.

★★★★★★★★★★★★★★★★★

It takes a *WISE* man to discover a *WISE* man.

... Diogenes

We can never judge another soul above the high water mark of our own.

WALK ON A RAINBOW TRAIL, walk on a trail of sun and all about you will be beauty. There is a way out of every dark mist over a rainbow trail.

Have you sighted your rainbow trail? Don't let it elude you. Grasp it and sing your *HAPPINESS*. You'll be the *WISER* if you do. Look through the sun and see through the mists. I see a rainbow above every cloud. You'll find *WISDOM* tied to the handle of the pot-of-gold. How much will be gained and retained depends on how hard you *"DIG."*

It is better to be small and shine than to be great and cast a shadow.

. . . Scrapbook

★★★★★★★★★★★★★★★★★★

A WISE duck always takes care of his bills first.

★★★★★★★★★★★★★★★★★★

Nine-tenths of *WISDOM* is being *WISE* in time.

. . .Dartnell Corporation

★★★★★★★★★★★★★★★★★★

God is the silent partner in all great enterprises.

> . . . Abraham Lincoln

It is a good thing to stop when you see red, whether you are in anger or in an auto.

> . . . Scrapbook

Ponder this. You could save a life, save a friend, save a *SOUL* or save a *SALE*. There is little doubt that the life, the friend and the soul are the most important but to save a *SALE* could be a stupendous stepping stone. The old adage of "counting to ten" would be *WISE* to heed, but *SAVING* or *SELLING, STOP BEFORE YOU PUSH THE ANGER BUTTON.*

We who work with *SELLING*, whether it be spiritual or temporal, it is important and expedient to observe the *"RED LIGHT."* The spiritual *SALESMAN* sells *LOVE, FAITH AND THE SAVING OF SOULS.* Anger is not a usable tool. The temporal *SALESMAN* has an equally hard task. You can not sell mere business deals, books or bread or any of the daily livables if anger is on the handle of the tool you use for digging. *WATCH* and *WAIT* for the *"RED LIGHT"* to warn you. *STOP, DON'T RUN IT.*

For most people, the hardest thing to give is in.

Mud thrown is ground lost.

There's a distinction between seeing the handwriting on the wall, and reading it.

It's easy to sit on the critics bench,
    And belittle the things men do.
Or give advise to those who failed,
    To carry their projects through;
Finding fault seems the stock-in-trade,
    Of critics who have never tried.
But I'll praise the man *WHO HAD A PLAN*,
    And whether it lived or died.
                    . . . Ira T. Peacock, Scrapbook

★★★★★★★★★★★★★★★★★

Would it not be better to taste your words before they pass your lips.

★★★★★★★★★★★★★★★★★

Beware—be sure your brain is engaged before putting your mouth in gear.
                    . . . Scrapbook

★★★★★★★★★★★★★★★★★

We give advice by the bucketful, but take it by the grain.
                    . . . William R. Alger

★★★★★★★★★★★★★★★★★

Every *WISE* man, whether builder, inventor, engineer, Doctor, teacher, writer or actor builds his *SUCCESS* largely on the accomplishments of others.

All philosophers and sages hold two things necessary to safely arrive at the knowledge of God's true *WISDOM:* First God's gracious guidance and second, man's assistance.

★★★★★★★★★★★★★★★★★★

A WISE FATHER SENDING HIS SON INTO THE WORLD SAID:

1. Tell the truth, falseheads are hard to remember.
2. Shine the heels of your shoes as well as the toes.
3. Don't lend money to your friends, you'll lose both.
4. Don't watch clocks, it will keep going, you do the same.
5. You do not always need clean cuffs, but you do need a clean conscience.
6. Don't borrow money unless you know where with all to pay it back. Then you don't need it.

. . . Scrapbook

★★★★★★★★★★★★★★★★★★

Every *WISE* man should keep a fair sized cemetery in which to bury the faults of his friends.

★★★★★★★★★★★★★★★★★★

A WISE old owl sat on an oak,
  The more he saw the less he spoke;
The less he spoke the more he heard;
  Why aren't we like that wise old bid?

★★★★★★★★★★★★★★★★★★

A Man always has two reasons for doing anything—a good reason and the real reason.

. . . J. P. Morgan

WISDOM always involves thought and proper action. *THINK FIRST—ACT AFTER THINKING.*

> . . . Garrett E. Case

Be a reader; this will help you become a leader.

THE WISE AND FOOLISH VIRGINS

1. Then shall the kingdom of heaven be likened unto ten virgins, which took their lamps, and went forth to meet the bridegroom.
2. And five of them were *WISE*, and five were foolish.
3. They that were foolish took their lamps, and took no oil with them;
4. But the *WISE* took oil in their vessels with their lamps.
5. While the bridegroom tarried, they all slumbered and slept.
6. And at midnight there was a cry made, Behold, the bridegroom cometh; get ye out to meet him.
7. Then all those virgins arose, and trimmed their lamps.
8. And the foolish said unto the *WISE*, give us of your oil; for our lamps are gone out.
9. But the *WISE* answered, saying, Not so; lest there be not enough for us and you: but go ye rather to them that sell and buy for yourselves.
10. And while they went to buy, the bridegroom came; and they that were ready went in with him to the marriage: and the door was shut.
11. Afterward came also the other virgins, saying, Lord, Lord, open to us.
12. But he answered and said, Verily, I say unto you, I know you not.
13. Watch therefore, for ye know neither the day nor the hour wherein the Son of man cometh.

To act with common sense, according to the moment, is the best *WISDOM* I know.

                                                    . . . Walpole

★★★★★★★★★★★★★★★★★

A warning is like an alarm clock. If you don't pay any heed to its ringing, some day it will go off and you won't hear it.

                                                    . . . Harris

★★★★★★★★★★★★★★★★★

Everyone who stops learning is old, whether this happens at twenty or sixty.   Anyone who keeps on learning not only remains young, but becomes constantly more valuable.

                                                    . . . Henry Ford

★★★★★★★★★★★★★★★★★

Time ripens all things, no man is born *WISE.*

                                                    . . . Cervantes

★★★★★★★★★★★★★★★★★

WISDOM is the fifth and last step on the ladder of *CHARACTER.* Whether it is acquired when young or old, it is a solid rung of the ladder. Ivor Griffeth once said, "*CHARACTER* is a victory, not a gift. It has to be earned." Are you at the top of the *LADDER?* Have you built your *CHARACTER* morally strong; strong enough to keep you from slipping on the top rung.

"Sow a thought you reap an act, sow and act and you reap a *CHARACTER.* Sow a *CHARACTER* and you reap a destiny." Have you opened your mind to all the opportunities and learned prudence and sagacity enough to handle your own destiny? *IS YOUR DESTINY SUCCESS OF FAILURE? IT IS IN YOUR HANDS!*

Failure

PART V

FAILURE OR?

A man can fail many times, but he isn't a real failure until he begins to blame someone else.

★★★★★★★★★★★★★★★★★★

When one hears the "never had a chance" yell of failure, think of Edison who never had a phonograph or an electric light bulb. Ford didn't have an automobile, the Wright brothers didn't have an aeroplane until they made one. You never had a chance until you make one.

... Coleman Cox

★★★★★★★★★★★★★★★★★★

Out of every 100 men called failures, 99% have been luke warm in their *WORK*.

★★★★★★★★★★★★★★★★★★

Any good caddy will tell you a golf ball isn't lost until you quit looking for it. By the same reasoning—a man isn't a failure until he *STOPS TRYING*.

... Scrapbook

★★★★★★★★★★★★★★★★★★

If you want to sit down in the middle of the road and say you are beaten, that it is of no use; there is little doubt that someone will oblige you and run right over you.

★★★★★★★★★★★★★★★★★★

They who are content to remain in the valley will get no news from the mountain.

★★★★★★★★★★★★★★★★★

An alibi is a "crutch" used by a weak sister.

★★★★★★★★★★★★★★★★★

Were *FAILURE* unknown how could one *SUCCEED*.

★★★★★★★★★★★★★★★★★

Learn to use todays defeats in a way that they become tomorrows victories.
★★★★★★★★★★★★★★★★★

Delay is responsible for more failures, more lack of *SUCCESS* than any other circumstance.

★★★★★★★★★★★★★★★★★

The man who tries using his hand at something and fails should try using his head for a change.

★★★★★★★★★★★★★★★★★

TEN EVERYDAY REASONS FOR FAILURE: Tardiness, lack of courtesy, big *I AM*, unwillingness to take advice, too dependent on advice, no self mastery, evil habits, fear and scatter-brained approach to *WORK*.

. . . Scrapbook

★★★★★★★★★★★★★★★★★

Many a *FAILURE* never gets his head above water because he never sticks his neck out.

Failing is not falling down, but remaining there when you have fallen.

★★★★★★★★★★★★★★★★★

It is hard to see why we are not *SUCCESSFUL,* but easy to see why our friends are a *FAILURE.*

★★★★★★★★★★★★★★★★★

Don't worry if you stumble and fall in life, the only thing that can't fall is a worm.

★★★★★★★★★★★★★★★★★

The man who falls down gets up lots faster than the man who lays down.

★★★★★★★★★★★★★★★★★

When you lose, weep softly, when you win, brag gently.

★★★★★★★★★★★★★★★★★

There are two kinds of *FAILURE:* those who thought and never did and those who did and never thought.

★★★★★★★★★★★★★★★★★

You may fail through laziness and fail from ignorance but the surest way to fail is a combination of both.

The only time you must not fail is the last time you try.

The man who wastes today lamenting yesterday will waste tomorrow lamenting today.

Henry Ford once said, "He who fears the future, fears *FAILURE* and limits his activities. *FAILURE* is only the opportunity to begin more intelligently again; there is a disgrace in fearing to *FAIL*."
His wealth and *SUCCESS* can be attributed to the following:
Because he had a will that would not break.
Because he had *PERSEVERANCE* that would not falter.
Because he trusted himself when all men doubted.

"Men attract to them the positions that belong to them. A small magnet never has the power to drag after it big things." *FAILURE* can be left behind and *SUCCESS* can be attracted by any one who uses a magnetic *PERSONALITY* and *PERSEVERANCE* to draw opportunity within his own reach.

★★★★★★★★★★★★★★★★★★

Success

PART VI

SUCCESS AND!

WHAT IS SUCCESS?
It is failure turned inside out.
The silver tints of the clouds of doubt.
And you can never tell how close you are,
It may be near and it may be far.
So stick to the fight when your hardest hit,
It's when things seem worst that you must not quit.

However things may seem, no evil thing is *SUCCESS* and no good thing is *FAILURE.*

. . . Longfellow

★★★★★★★★★★★★★★★★★

The *FATHER* of *SUCCESS* is *WORK.*
The *MOTHER* is *AMBITION.*
The oldest son is *COMMON SENSE.* Some of the boys are *HONESTY,   PERSEVERANCE,   THOROUGHNESS, FORESIGHT, ENTHUSIASM* and *COOPERATION.* The oldest daughter is *CHARACTER.* Some sisters are *CHEER-FULNESS, LOYALTY, COURTESY, CARE* and *ECONOMY, SINCERITY* and *HARMONY.*
The baby is *OPPORTUNITY.* Get acquainted with the *"OLD MAN"* and you will get along with the rest of the family very well.

. . . Scrapbook

SUCCESS is never found at the top of the hill if the duties at the foot are neglected.

★★★★★★★★★★★★★★★★★★

Thorough preparation is behind every great *SUCCESS.*

★★★★★★★★★★★★★★★★★★

MEN WHO SUCCEED: Correct their own faults, do not expect pay for half a weeks *WORK*, do it today—not—tomorrow, cooperate with other men, are loyal to themselves and their employers, and study constant to prepare for a higher position.

★★★★★★★★★★★★★★★★★★

EACH IS GIVEN A BAG OF TOOLS,
   A shapeless mass, a book of rules,
And each must make 'ere life has flown
   A stumbling block or a stepping stone.

★★★★★★★★★★★★★★★★★★

Play the perfect ball game. Touch all bases. Don't leave out planning to be the player who wants to score but never gets to first base.

. . . Scrapbook

★★★★★★★★★★★★★★★★★★

FORMULA FOR SUCCESS:
"Planning *X* Perspiration *X* Personal *WORK X* Persistence *X* Presentation *X* Personality, = *PERFECT PERFORMANCE.*"

★★★★★★★★★★★★★★★★★★

I contend that the scientist, business man or thinker who gave the world the *FORMULA FOR SUCCESS* has indeed multiplied his special talents. *PRACTICE WILL PAY OFF IN PERFECT PERFORMANCE.*

★★★★★★★★★★★★★★★★★

The discipline and self control of Jesus would have made him *SUCCESSFUL* in any *UNDERTAKING*. Small wonder, I think, that we are taught by our leaders to follow his admonition.

★★★★★★★★★★★★★★★★★

Alice In Wonderland was found searching for the strange lock among the bushes. "What are you searching for," she was asked. "The key to unlock myself, if I do not find it I am doomed to remain shut up within myself forever." This bit of advise I urge you to consider. *UNLOCK YOUR GOD GIVEN POTENTIAL.* Your faith in yourself is the key to open the doors of your world.

★★★★★★★★★★★★★★★★★

TO DEVELOP SUCCESS: Always keep your promises. Always be on time. Never forsake right to follow the line of least resistance. It is far better to do things the right way than the wrong. It is smart to be honest—stupid to be dishonest.

★★★★★★★★★★★★★★★★★

Shut yourself in a dark room and in time you will go blind. Close your eyes to the opportunities about you and look only on the dark side of life and you will be so blind that you will never see *SUCCESS.*

. . . Scrapbook

There is no greater obstacle in the way of *SUCCESS* in life than trusting for something to *TURN UP*, instead of going to *WORK* and *TURNING SOMETHING UP*.

★★★★★★★★★★★★★★★★★

A reasonable amount of intelligence and a lot of footwork will make a *SUCCESS* of anyone.

★★★★★★★★★★★★★★★★★

IT CAN BE DONE

If you can't be a pine on the top of the hill,
  Be a scrub in the valley—but be
The best little scrub by the side of the rill;
  . Be a bush if you can't be a tree.
If you can't be a bush, be a bit of the grass,
  And some highway some happier make;
If you can't be the muskie, then just be a bass-
  But the liveliest bass in the lake!
If you can't be a highway, then just be a trail;
  If you can't be the sun, be a star,
It isn't by size that you win or you fail—
  Be the best of whatever you are.

. . . Scrapbook

★★★★★★★★★★★★★★★★★

AS A RULE: It is the sternest philosophy, but truest that in the side arena of the world, *FAILURES* and *SUCCESS* are not accident but the strictest justice. If you do your fair days *WORK* you're certain to get a fair days wages—in praise or pudding. Into the world a man brings his *PERSONALITY* and his biography is simply a catalogue of its results.

. . . Alexander Smith

The most important single ingredient in the formula of *SUCCESS* is the knack of getting along with people.

. . . Theodore Roosevelt

★★★★★★★★★★★★★★★★★★

ONE OF THE MOST IMPORTANT LESSONS IN LIFE is that *SUCCESS* must continually be won and is never finally achieved . . . Those who achieve fame, fortune and influence only gain another level of responsibility in which they must make good.

★★★★★★★★★★★★★★★★★★

A retired business executive was asked his secret to *SUCCESS*, he replied, "it could be summed up in three words and then some. I discovered at the early age, that most of the difference between average and top people could be explained in three words. The top people did what was expected of them—*AND THEN SOME*.

★★★★★★★★★★★★★★★★★★

The fellow who "gets there" is the one who uses his *FAILURES* as 'Stepping stones" instead of grave stones.

★★★★★★★★★★★★★★★★★★

Genius is only the power of continuous effort. The line between *SUCCESS* and failure is so fine we scarcely know when we pass it. Many a man has thrown up his hands at a time when a little more effort, a little more patience would have achieved *SUCCESS*. There is no failure except in no longer trying.

. . . Elbert Hubbard

★★★★★★★★★★★★★★★★★★

The light hidden beneath a bushel basket causes no radiance. *GOLD* is valueless unless *labor digs* it and puts it to use. The diamond is worthless in its native clay. Of what worth is a pearl lying at the bottom of the ocean bed. *SO IT IS WITH ABILITY.* If not actively and properly used, it has no place in the *SUCCESS OF MAN.*

. . . Scrapbook

★★★★★★★★★★★★★★★★★

One thing better than *SUCCESS* is to be worthy of *SUCCESS.*

★★★★★★★★★★★★★★★★★

The difference between *SUCCESS* and *FAILURE* is measured by one's ability to concentrate.

★★★★★★★★★★★★★★★★★

You will find the key to *SUCCESS* under the alarm clock.

★★★★★★★★★★★★★★★★★

SUCCESS comes more easily to some people than to others.

★★★★★★★★★★★★★★★★★

Whenever an individual or a business *SUCCESS* has been attained, progress stops.

★★★★★★★★★★★★★★★★★

Behind every *SUCCESSFUL* man is a man who says he went to school with him.

★★★★★★★★★★★★★★★★★

1. *THINK* straight and you will act straight.
2. *ANALYZE* things, get all the facts before concluding.
3. *DEVELOP* habits of cleanliness and orderliness.
4. *SET* a reasonable goal and determine to reach it.
5. *TAKE* advise but do your own thinking.
6. *CHEER* up the other fellow, but keep your own troubles to yourself.
7. *NEVER* admit to anyone, even yourself, that you are licked.
8. *SPEND* a little less than you have earned.
9. *MAKE* friends, but remember, the best wear out if you use them.
10. *DON'T* be afraid to dream, a little dreaming and imagination are necessary for *SUCCESS*.

. . . Scrapbook

★★★★★★★★★★★★★★★★★★

These ten simple but material rules do not include wealth and property, but young or old, the person who follows them will uncover a glorious mine of *SUCCESS AND HAPPINESS!!!*

★★★★★★★★★★★★★★★★★

THE SCALES OF SUCCESS

4 oz. *HARD WORK*	versus	4 Oz. *IDLENESS*
4 oz. *SKILLED EXPERIENCE*	versus	4 Oz. *INEXPERIENCE*
4 oz. *GOOD CHARACTER*	versus	4 Oz. *IRRESPONSIBILITY*
4 oz. *HAPPINESS & LOVE*	versus	4 Oz. *MISERY & FEAR*

16 oz. = lb. *SUCCESS*   16 Oz. = 1 lb. *FAILURE*

Can you tip the needle and balance my *SUCCESS SCALES* to bring you the achievements necessary to fill your *"POT OF GOLD."*

PART VII

HAPPINESS!!!

Some people try to purchase *HAPPINESS* by having riches and fame, but without *SERVICE TO ALL* Mankind there can be no *HAPPINESS*. One man said, "To try to win *HAPPINESS* from the world without serving is like trying to distill gasoline from water instead of from crude oil. It can't be done."

★★★★★★★★★★★★★★★★★

The statesman who forwards the cause of humanity wins *HAPPINESS*.
The judge who reads the law in the light of common sense wins *HAPPINESS*.
The law of *HAPPINESS* is as inexorable as the law of gravitation. This is a tremendous law.

★★★★★★★★★★★★★★★★★

HAPPINESS is a mosaic, complexed of many smaller stones. It is an act of kindness, the disposition to accommodate, to be helpful, to be sympathetic, to be unselfish, to be charitable, to be considerate—these are the little things which added up at night, are found to be the secret of a happy day.

★★★★★★★★★★★★★★★★★

HAPPINESS in this world when it comes, comes incidentally. Make it the object of pursuit and it leads us on a wild goose chase and it is never attained.

★★★★★★★★★★★★★★★★★

Our hours are loans from *OLD FATHER TIME.* Are they paying you dividends of prosperity as well as happiness?

Every time we laugh, we take a kink out of the chain of life.

★★★★★★★★★★★★★★★★★

Friendly people make a friendly company. Don't be a sour apple.

★★★★★★★★★★★★★★★★★

Two persons will not be friends long if they cannot forgive each others failure.

★★★★★★★★★★★★★★★★★

No man is rich enough to be without a neighbor.

★★★★★★★★★★★★★★★★★

So long as we love we serve; so long as we are loved by others, we are indispensable. No man can be useless who has a friend.

★★★★★★★★★★★★★★★★★

HAPPINESS is like jam—you can't spread even a little without getting some on yourself.

★★★★★★★★★★★★★★★★★

True *HAPPINESS* consists not in the multitude of friends, but in the worth and choice.

★★★★★★★★★★★★★★★★★

If you want to spend a happy twilight in peace and solitude, Be prepared to do all the good you can, to all the people you can, wherever you can, *WHILE YOU CAN*, and pray that when you land in a "rocking chair," someone as helpful, can rock you there.   For me, "daylight" is the time to begin, *WHILE I CAN.*

★★★★★★★★★★★★★★★★★★·

HAPPINESS is a state of the mind that results from banishment of sorrow, worry and fear and in their place there is the aspiration to accomplish that which gives joy and peace to the heart, mind and soul.

★★★★★★★★★★★★★★★★★

THREE FACETS OF HAPPINESS
1. The finest science—extracting sunshine from a cloudy day.
2. The finest music—*LAUGHTER.*
3. The finest law—*THE GOLDEN RULE.*

★★★★★★★★★★★★★★

RULES FOR HAPPINESS
RULE YOUR MOODS: cultivate a mental attitude of peace and good will.

BE GRATEFUL: begin each day with gratitude for your opportunities and blessings.

BE GLAD: for the privilege of life and good *WORK.*

BE INTERESTED: in others and direct your mind from self-centeredness. In the degree that you give, serve and help, you will experience the by-product of *HAPPINESS.*

CULTIVATE: a yielding disposition.   Resist vanity and see others points of view.

GIVE GENEROUSLY: of your time and talents.

★★★★★★★★★★★★★★★★★★★

When the sun in the skies is blotted out by fear, worry and despair, depend on the inner sun within you. Generate sunshine and *HAPPINESS* wherever you go. Light the way through darkness and defeat.

★★★★★★★★★★★★★★★★★★★

How to crate this inner sunshine? Close your mind to negative thoughts as you would lock the doors to robers. Fill your mind with hope, cheer and positive things. Associate with men of great faith, men who believe things can be done and who roll up their sleeves and do for them.

. . . Scrapbook

★★★★★★★★★★★★★★★★★★

He who carries his sunshine with him has learned the master secret of triumphant living. Work with me, day by day in weaving a *"MAGIC BASKET IN WHICH WE CAN CARRY OUR SUNSHINE IN.*

★★★★★★★★★★★★★★★★★★

WHAT THE WORLD NEEDS NOW: is a rebirth of *FAITH.* The world will not move without *FAITH.* If you had no faith in the postman, you would not post a letter in the box. If you did not have faith in the alarm clock, you would not wind it. If you did not have *FAITH* in your lock, you would leave your door open. In short, when you do not have *FAITH,* you discard things as of no value. You *CANNOT SUCCEED* in any job without *FAITH* in your customers, your company, your branch, but greatest of all, *YOU MUST HAVE FAITH IN YOURSELF IN YOURSELF.*

The future *SUCCESS* and *HAPPINESS* of an individual, a family, a business and a nation is built on *FAITH*. So long as this *FAITH* is strong, history shows that no trials are too hard to overcome. Only when faith weakens, then do people lose the ability to master their circumstance.

The basis of all *FAITH* is *FAITH* in one's self. Our country has been made great by men and women who had *FAITH* in themselves.

The Declaration of Independence and the Constitution were supreme *ACTS OF FAITH*. Our industries have made life in the United States safer, fuller and happier than anywhere in the world, because men of vision and courage have adventured into great exciting fields.

Today there are signs that this *FAITH* is fading. What we need is renewed *FAITH*. We need courage and confidence. We need *FAITH* in those around us in business, church, schools and in the country which is still the world's greatest land of opportunity. But above all we need to renew our *FAITH* in *GOD* and ourselves.

. . . Scrapbook

This article would have been written at least nine years ago this 1975. What foresight this person must have had. No author's name was printed but if he or she should be known, he should be fairly praised for this inspirational and futuristic bit of *WISDOM*.

ANYONE, be he a Doctor, lawyer, merchant, chief, can dig in the mine of *HAPPINESS* and *FAITH* and can put his or her foot on the top rung of the *SUCCESS LADDER*. However, do not be over confident, bear in mind that "Lifes Ladder is full of splinters and hurts most when coming down."

★★★★★★★★★★★★★★★★★

Having read and given thought to these gems of *GOLD*, let not this be the end; but the beginning. Ask yourself, "Am I a *SUCCESS?* Have I found *HAPPINESS* or am I still searching and digging for the *"POT-OF-GOLD"* at the end of the rainbow.

★★★★★★★★★★★★★★★★★

WHEN YOU CATCH IT, WORK TO HANG ON TO IT and God grant that you may keep it. Whatever your *SUCCESS*, keep these two tools, *FAITH* and *PRAYER* with you. *"LET PRAYER BE YOUR KEY OF THE DAY AND YOUR LOCK OF THE NIGHT."*

★★★★★★★★★★★★★★★★★

"BEGIN AT HOME."   Be *SUCCESSFUL* and solvent in your own business and professions.

SERVE your communities—your church—your country— and many good causes.

REACH out worldwide, and do something for someone, somewhere.

. . . Richard L. Evans

★★★★★★★★★★★★★★★★★

"Seek ye first the kingdom of God and his righteousness; and all these things shall be added unto you.

. . . Matt. 6-33

★★★★★★★★★★★★★★★★★

A sincere thanks to all authors and readers who have given me the *FAITH* to *CHALLENGE* you to discover the mine of *SUCCESS* and *"DIG YOUR OWN GOLD."*

Written and compiled by
Ellen Gallup Genta

EPILOGUE
Written after June 5th, 1976
(After Teton Dam broke and flooded Eastern Idaho)

GOD CREATED THE EARTH TO ENJOY, NOT FOR MAN TO DESTROY

"WHATEVER COMES, this too shall pass away."

(Ella Wheeler Wilcox,)

IT CAME! The Teton flood, ravishing and roiling and cutting down the good earth of God. Of the many millions of prized family treasures that were destroyed; one was my *DAD'S SCRAPBOOK*. It did not float away as so many dearly beloved possessions did, but very little of it is readable. Thank God, that earlier, I was wise enough to write *"DIG YOUR OWN GOLD."*

There are no words, no writing, no picture, however fluent or vivid, that can show the world the caldron of chaos, the apathetic trance and the empty bewilderment that is mirrored in these Idaho communities. Only those who saw it, believe it! Only those who see the faces of the homeless fathers and mothers, the widows and the old ones who have worked a life time and lost that lifetimes in a few hours; only those, can believe that so much misery and destruction abides. Anger, bafflement and defeat followed the complete disaster that devastated this valley of homes, of loved ones, of productive farms and precious animals which ensured a livelihood.

I have read and reread the following words of Richard L. Evans, the beloved sage, Perhaps, with Faith in God, they will bring comfort to all who are sad and weary and who have lost family remembrances which can never be replaced by any amount of wealth.

("No man is himself in acute sorrow. No man is himself in anger. No man is himself with feelings of offense. And decisions that will wait are safer with waiting—waiting for time to take over, for the dust to clear away, for tempers to cool, for prespective to return, for the real issues to show themselves, for the real values to reappear, for judgement to emerge and mature.

We should think seriously before we slam doors, before we burn bridges, before we saw off the limb on which we find ourselves sitting. Decisions in acute sorrow, decisions in anger, decisions under pressure, decisions that haven't been thought through are less likely to be mature and safe decisions.")

He also said, ("There is no loneliness so great, so absolute, so utterly complete, as the loneliness of a man who cannot call upon his God.)

IN THE BEGINNING

GOD CREATED THE EARTH TO ENJOY, NOT FOR MAN TO DESTROY

"The earth is the Lords and the fullness thereof; the world,
    and they that dwell therein,"
Were intended to share together the beauty and the love lodged within.
God created the good earth and he surely meant,
That man live in love and peace, happiness and content.
Man was given his freedom as a gift of love,
To work and to labor and secure his blessings from above.
But man soon destroyed the good land of God,
By building a dam that ravished our sod.
A small crack appeared in the bulwark, the crevice became
    bold,
Too large for a little boy's thumb to let the dike hold.

The water surged and roared, ugly with mud and terrain,
It cut down the forest with ruthless disdain.
Earth's natural beauty was a channel of chaos, a horrible hell.
Our paradise was destroyed by a madness that no hand could
  quell. . .

In the beginning, God created the earth for man to enjoy,
But he could not stop man from the mistakes they employ.
For nothing on earth is forever, and what earthly good
  we possess,
Can be corrupted by flood and fire into complete nothingness.
So lay up treasures of friendship that will last forever,
These are the riches you can hold and will part with never.
In the deep dark hours of rebuilding this miserable mess,
Trust in the Lord, bear your cross and accept this distress.
We are a staunch people and on recovery we're bent,
Man must learn that in whatsoever state he is in, there with
  to be content.
Don't go it alone, talk it over with God,
Seek his hand in rebuilding the land you have trod.
"For no man is an island," no man need be alone,
Who has neighbors and friends whose brotherhood has shown.

This valley is made of brave people, the most courageous on
  earth,
Ask the merciful helpers who came, they can vouch for their
  worth.
To the world they will show what real people can do,
When faced with destruction and adversity to.
I'm proud, mighty proud, to be one of such spirited
  population
Who held on to their sanity through such a sad desolation
One little lady cried, as the tears splashed her tired face.
"In my lifetime, I've never had to ask for anything, from
  anyplace.
So we all put pride in our pockets, when you came from afar,
THANK YOU, GOD BLESS YOU WHOEVER YOU ARE!

*FOR LIGHTENING OUR BURDEN ALMOST TO HEAVY TO
ENDURE.*

                                        . . . E.E.G.

DO YOU BELIEVE?

Do you believe in miracles, in faith, hope and prayer?
    Do you believe in these futile wars, that God is really there?
When our earth and it's elements are a trembling mass
    And a flood rages madly through forest and grass,
And you hear the moans of the hungry hordes
    Of beasts and men who have laid down their swords;
Unable to fend for life and plead their own cause
    With agressors who unjustly transgressed God's laws
Of loving they neighbor like the Good Samaritan?
    Small wonder indeed, this corruption of man!

Do you believe in miracles, faith, hope and prayer?
    Do you believe that God and his mercy can really be there?
I believe in this modern day and tumultuous clime
    Man should look for God in his heart and find his own time
To see beauty and love, and find a purpose in care,
    And give hope, faith and love to comrades wrought in
despair.
I believe in the miracle that you can find in a star,
    A rose or a rainbow, where ever you are.
Each one must find his destiny in his own heart and way,
    For no man can shoulder the burdens of a torn world today.

WHAT THE WORLD NEEDS NOW is a rebirth of *FAITH*,
    And to love thy neighbor, thus the Lord saith.
'Twould start miracles rolling and they would gather love's
moss.
    It would be a rebirth for He, who hung on the cross.
Do you believe in yourself, in your comrades and this nation?
    Do you believe that God's love could bring you exhaltation?
Do you believe in miracles and faith, hope and prayer?
    Do you believe that you'll find happiness, if you but share
One bit of sunshine, a bit of food or a flower
    With a friend who needs love's wisdom, courage and power.
I DO . . . I BELIEVE IN MIRACLES!

                              . . . E.G.G.